I0235970

A SIMPLE
Systems Engineering
Guide for Success

DAYMOND E. LAVINE

Omniversal Publishing House
2124 Farrington Street, Suite 350
Dallas, TX 75207

Omniversal Publishing House is a subsidiary of
Moksha Media, moksha-media.com

Copyright© 2017 by Daymond E. Lavine, Daymond & Co., daymondco.com

Library of Congress Control Number: 2017909240

ISBN-10: 0-9961323-3-3 / ISBN-13: 978-0-9961323-3-6

E-book ISBN-10: 0-9961323-4-1 / ISBN-13: 978-0-9961323-4-3

Book Cover Design by Daymond E. Lavine of Moksha Media, moksha-media.com

Printed in USA by Lightning Source

10 9 8 7 6 5 4 3 2 1

All rights reserved. Without limiting the rights under copyrights reserved above, no part of this publication may be reproduced, stored in or introduced into a retrieval system, or transmitted, in any form or by any means (electronic, mechanical, photocopying, scanning, video presentation, audio recording, or otherwise), without the prior written permission of both the copyright owners and the above publisher of this book.

For additional information about the author, to book a signing event, or for information on acquiring written permission in the case of brief quotations embodied in critical articles, interviews and reviews, visit www.DaymondCo.com.

©*Daymond E. Lavine*

Publisher's Note

The author and publisher of this work intend for this work to provide general information about the subject matter covered. This work is not intended to provide legal opinions, qualify advice, or serve as a substitute for advice by licensed, legal professionals. This work is sold with the understanding that the author and the publisher of this work are not engaged in rendering legal or other professional services.

The author and publisher of this work do not warrant that this work is complete or accurate, and do not assume and hereby disclaims any liability to any person for any loss or damage caused by errors, inaccuracies or omissions, or usage of the information herein.

Dedications

To Mrs. Patsy Bazile, my loving, generous, and beautiful mother. Her heart is so large and forgiving that it pours out kindness sometimes too much to everyone else around her except for herself. That's okay; I remind her about it constantly. This woman is one of faith and creativity, with treasures so profound they shimmer gold hues whenever anyone is in her presence. May God continue to bless and adorn her. I love you, mom!

To Mr. Dezria Bazile, my honorable stepfather, a man with a cool exterior that never quite shows exactly who he is inside. But actions always speak louder than words, and he has been the only father I've ever really known. Thank you, Mr. B., for stepping in and being responsible in ways you did not have to be. That speaks infinite volumes for me.

To Stanley Coleman, my phenomenally supportive partner, with a heart as deep as the deepest ocean. He is that constant rock in my life who has been willing to see things to the very end with me. He has never failed to demonstrate his eager desire to team and grow with me. Thank you so much, Stanley. At this point, I think it's safe to say, nothing can tear down the empire now!

To Mr. Ted Martens, the kindest and funniest mentor I have ever known who is just as smart as he is hilarious! He is the friend I found in that huge world I entered called Systems Engineering. He actually made it place I grew to like and became comfortable in. Thank you, Ted! May your retirement years pay back all those countless mentoring hours you devoted to so many engineers who are much better because of it!

©Daymond E. Lavine

Author's Note

In my youth and all throughout my adolescence, I had never really been impressed with myself. Today I guess I'm alright! (I'm smiling.) As many of us do throughout those very early years of our lives, I spent a lot of time trying to figure out people and situations. I wondered where I fit into everything—this very confusing and often misunderstood stew of incidents strung together in what we simply call "life." As I struggled to gain my sense of self, I became my own worst critic. In fact, on many days, I even became my own worst enemy. To this very day, I cannot quite understand where my tendencies for perfectionism and people-pleasing came from. However, I must admit that these qualities of mine were the same ones that made me who I am today. Through lots of prayer, self-introspection, and willingness to thrive on being my personal best, I have become keenly aware of how I fit into my current situation. One of my primary purposes in life is inspiring others. I am especially dedicated to helping people see how they can become successful beyond what they perceive to be their own limitations.

So, what is success? Everyone has different ideas about what it is. Some people believe that success equates to lots of fame and fortune. Others believe that success equates to minimalism and happiness. There are an infinite number of other definitions that exist between these two extremes, but it is safe to say that success is totally subjective. This is a very simple truth that many people forget. As a result, they spend a great amount of time and effort trying to become "successful" because they chase a moving target. They attempt to acquire status, possessions, and experiences that make them *feel as though they are successful* for fleeting moments of time. This can be a very miserable existence for people who have not figured out what success means specifically to them, and I am glad I learned this valuable lesson in life early on.

As I exited my college years, I began looking inside myself for answers to profound questions such as "Who am I?" and "What is my purpose in life?" I questioned my value system and the value systems of others around me. I even had struggles coming to terms with my spiritual and religious beliefs. When I think back on those years, I know now they were turning points that clearly defined who I am.

As a child, I had always been creative. I used to doodle in class on pieces of paper, sketching caricatures and images of people and objects. I helped my mother complete various creative projects for church functions that she organized and managed, such as plays and social events. I participated in theme poster contests in my grade school years. In my high school years, I sketched and painted football game burst-through posters. In my senior year, I even participated in the year book committee and laid out the interior of the yearbook. Still, I balanced all that creativity with maintaining a 4.0+ grade point average (GPA). In fact, when I graduated from high school, my cumulative GPA was 4.125, and I was identified as the graduating class Salutatorian, second to the Valedictorian. So, my mind was creatively and technically balanced. I actually loved my math classes, and Calculus was a breeze for me. My balanced mind was both a blessing and a curse. I did not have a solid notion of what exactly I wanted to be when I grew up. I had no preferred career in mind because there were a multitude of directions I could go in. Thus, I remember making the decision at my young age to choose a career in law, computer science, engineering, or medicine. I did not believe at that time creative jobs would be lucrative. I wish that in the early 90s, I would have had a crystal ball that showed me the word "branding" would take on a monster of a meaning in 2000 and beyond. Hindsight is 20/20. Now, I offer those services through a business entity of mine. Nonetheless, I initially began my college classes majoring in Computer Science when I started attending Xavier University of Louisiana in 1995. During my second semester, I changed my major to Dual-degree Engineering because I liked math much more than Computer Science. I also thought it would be great to obtain a degree in Physics as well as Electrical Engineering in a unique program offered by Xavier University in association with other partnering universities. My changed curriculum proved to be very challenging, but it was not my only challenge in college. I also had to come to terms with my spiritually. So, I was not only sharpening my technical depth; I learned to expand my spiritual depth as well.

I grew up attending and participating in the church. I took part in many church functions, and listened very intently to the preachers during all the church services I attended. I loved listening to all the joyful gospel songs and hymns that sounded throughout the walls of the building, and I continued going to church when I got to college. As part of my college curriculum, theology classes at Xavier University of Louisiana were mandatory. The school is a private, Catholic university. Thus, in those classes, at a high level, I learned about the many religions that exist. I also learned about the life of Jesus Christ. This was a profound turning point for me. It was then that Jesus became very *real* for me, despite all the church services I had attended. I learned that Jesus was not elitist, prejudiced, or demeaning. You see, the way that I had perceived the messages preached to me in the church made me feel as though He was an untouchable, historical figure that I would never have been able to interact with. However, in my theology classes I learned that Jesus Christ did not separate Himself from wrong-doers like I had imagined He did based on all those church sermons I heard. In my mind, I thought He had existed on this extremely high pedestal that no one could ever aspire to reach.

By the time I got half way though my theology classes, I found myself challenging all my beliefs, and "something" or "someone" took notice. I experienced the worst semester I ever had. My car was vandalized twice. The first time, equipment was stolen. The second time, school funds were stolen from my glove compartment. It was money that I needed to deposit in the bank that was supposed to be used for my living expenses. Afterwards, during that semester, I struggled financially to pay bills and other living expenses. I even had two flat tires, which in turn, forced me to tap into savings to cover those unexpected expenses. I remember questioning myself one day, "Why is all of this happening to me?" The answer that came back was quite simple: "All of this is happening to you because you are doubting your beliefs." I cannot tell you that I actually heard the voice bellow out of thin air. However, I can tell you that the answer came to me as clear as day. I *heard* it from within. It was then I learned how to *converse* with God. I realized at that moment I did not need to go anywhere or do anything special for God to respond to me. I simply needed to call out and question myself and Him about a situation I was going through. Then a response would come back to me—a response from something much bigger than myself. That was the impetus of my recognition of the connectivity of everything—events, people, circumstances, and God.

I went on to graduate from Xavier University of Louisiana and the University of New Orleans with Bachelor's degrees in both Physics and Electrical Engineering. I worked in the field of nonprofit immediately after college; but soon after, I landed a Systems Engineering job with a leading government contracting company. I boldly marched into my adult life aiming for success while developing my technical career. At the same time, I continued to work on creative side projects while exercising my spiritually. I was becoming a "success"! Yet, I was not conscientiously paving a path toward my happiness.

I am extremely happy now that I recognize the connectivity of it all. I do not think my writing of this book is a fluke. I have a creative mind and technical expertise that benefit me daily. I have a very strong connection to God with a keen sense of how every human being has something to offer to the Universe. To round it all out, I have had a successful Systems Engineering career, which has not only allowed me to thrive in the work place, but also strategically help others through my other businesses and services. My Systems Engineering background has allowed me to treat "success" as a total system, organized as strategic objectives, tasks, and measures.

I am writing this book because I have had many conversations with friends, clients, and colleagues that wonder how I "do it all." I know that I exist to inspire others as one of my many purposes in life. This is a book of inspiration and a success guide for others to use for implementing some powerful Systems Engineering techniques I have learned to apply in my daily life and toward my creative projects. I have, to this very day, remained creatively driven, but I am also technically inclined. I promise you I have made this book easy to understand. Do not be afraid to read further. I have only focused on the key aspects of an engineering model that anyone can use to be successful at any number of projects and personal endeavors. I hope that you like reading this book. Even if it is not the most enjoyable experience you will ever have, I

©Daymond E. Lavine

definitely know, for sure, that you will learn a thing or two. I know that you will be able to use the methodology I have identified herein for your personal success. Happy reading!

Table of Contents

Introduction **12**

Figure 1. The Success System V^{SM} 15

Section I **18**

Part 1
Conceive Your Success **20**

Success Tip #1 22

Part 2
Plan Your Success **26**

Success Tip #2 28
Success Tip #3 29
Success Tip #4 34
Success Tip #5 36
Success Tip #6 40
Success Tip #7 42
Success Tip #8 45
Success Tip #9 48
Figure 2. Simple Success Schedule 49
Figure 3. Simple Success Budget 50
Success Tip #10 52

Part 3
Define Your Success **55**

Success Tip #11 57
Success Tip #12 60
Success Tip #13 66
Success Tip #14 68

Part 4
Build Your Success 71

Success Tip #15 73
Figure 4. Success System V^{SM} Left-side Flow Down.................................76

Section II 78

Part 1
Test Your Success 80

Success Tip #16 82
Success Tip #17 85

Part 2
Verify Your Success 86

Success Tip #18 88

Part 3
Validate Your Success 91

Success Tip #19 93

Part 4
Actualize Your Success... 94

Success Tip #20 96
Success Tip #21 98

Conclusion.. 99

Introduction

©Daymond E. Lavine

So here comes the boring part: I must tell you exactly what Systems Engineering is. In the past, when I would mention to people I was a Systems Engineer, the first thing that would pop into their heads was that I did computer science work. If it was not that, then they thought I did something related to computers. So, I would clarify to them I only used computers to do my Systems Engineering work. Then, I would tell them that as a Systems Engineer, I would act more as a liaison in my everyday job who worked among the many other engineering disciplines in my organization. My job was to ensure that the Systems we built came together in accordance with the specifications that were defined for them. The other people I worked with included Mechanical Engineers, Electrical Engineers, Software Engineers, Safety Engineers, and a host of other applicable personnel according to the system being built.

I knew that the people to whom I explained my job still were not clear, but there was really no precise way of conveying to them what I did in a casual conversation. The truth is, even as I write this book, I will not be able to give you a true sense of what I did as a Systems Engineer. My job varied quite a bit from project to project. However, thankfully, I have developed a sound strategy for success, which precipitated out of those many years. Now, I can share it with you! It will not help you completely understand the field of Systems Engineering, but it will help you employ a rigorous methodology, based on Systems Engineering concepts, that helps you take your successes in life to new heights.

A plain and simple definition for Systems Engineering is that it is the multidisciplinary management of the design, or creation, of a complex system from conception to actualization. In other words, you know that you want something to exist, so you conceive it; and then, you realize it using various techniques, strategic plans, and standard practices. Believe it or not, there is a standard Systems Engineering model that helps all Systems Engineers do their very complex jobs. Depending on where a system is in its development lifecycle, which could last for years, Systems Engineers do different tasks to meet the need of the System at specific times.

All this seems very complicated, right? Well, that is why we needed a work process model that we called the "Systems Engineering V." After approximately three solid years of gaining some fantastic experiences as a Systems Engineer, I quickly learned that almost anything could be perceived as a "System." That meant I could use some of those very same Systems Engineering techniques I used in my corporate job to get tasks accomplished for my entrepreneurial projects. I could even use them to overcome the many challenges I faced in my everyday life. There is no need for me to show you the Systems Engineering V here. If you are interested in knowing more about it, just Google it. There is plenty of information about it online. However, what I will share with you is my **Success System V**[SM]. This V-model, along with the success generation methodology I provide herein, is specifically created to help you succeed! We will need it as we get through the rest of this book, and it is provided for your benefit on the following page. This is the time you should mark or note the page depicting the Success System V[SM] for later reference. This book is entirely arranged based on that model. We will travel the Success System V[SM] from left to right, down the left side and up the right side.

©Daymond E. Lavine

CONCEIVE YOUR SUCCESS

ACTUALIZE YOUR SUCCESS

PLAN YOUR SUCCESS

VALIDATE YOUR SUCCESS

Problem Definition & Decomposition

DEFINE YOUR SUCCESS

VERIFY YOUR SUCCESS

Problem Assessment & Minimization

BUILD YOUR SUCCESS

TEST YOUR SUCCESS

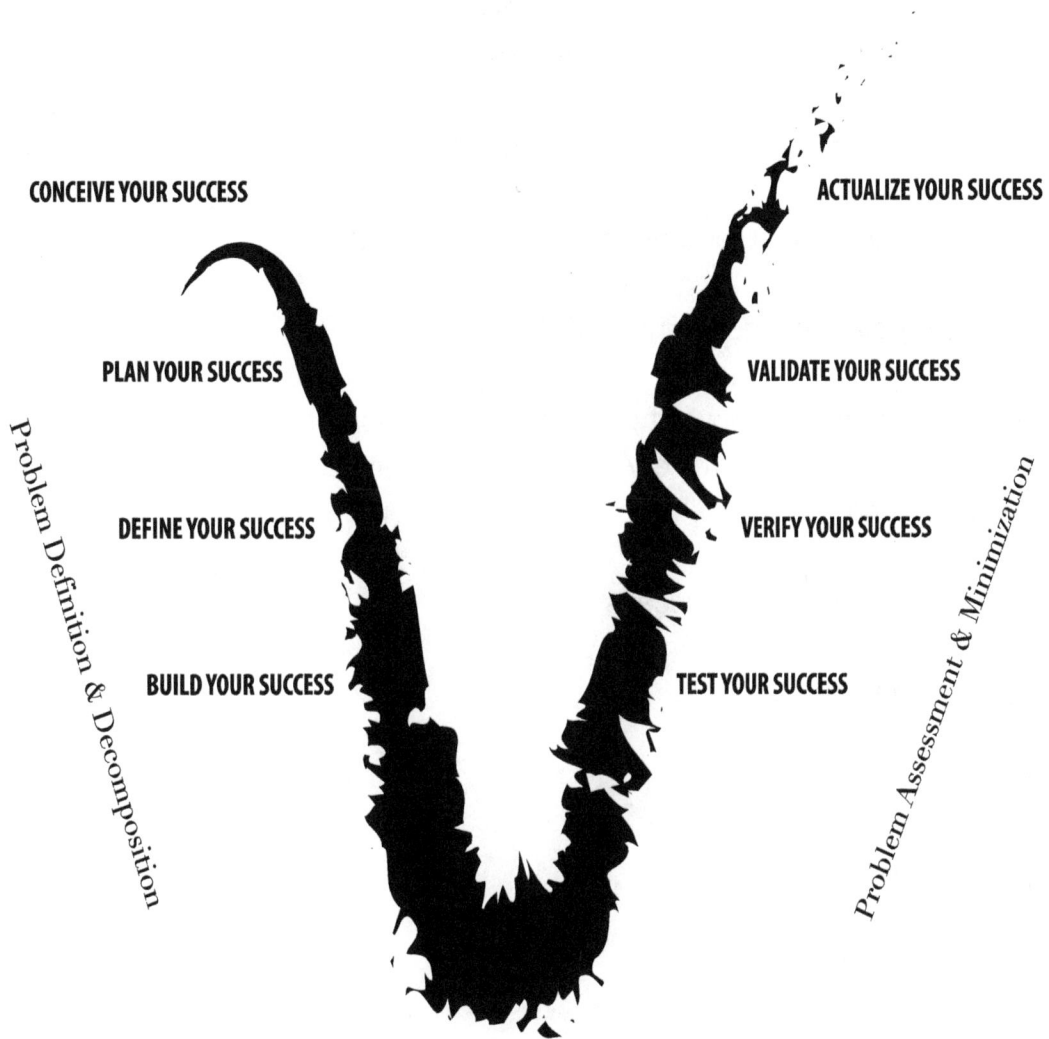

Solution Set Generation & Implementation

Figure 1. The Success System V[SM]

My Success System V^{SM} organizes the attainment of success into three primary stages as depicted in **Figure 1**. Those stages are the following:

(1)   Problem Definition and Decomposition

(2)   Solution Set Generation and Implementation

(3)   Problem Assessment and Minimization

As a person makes his or her way through each of these primary stages, he or she will have the opportunity to perform very specific tasks that help with the achievement of the success he or she is seeking. Those tasks, in chronological order, are the following:

(1)   Conceive Your Success

(2)   Plan Your Success

(3)   Define Your Success

(4)   Build Your Success

(5)   Test Your Success

(6)   Verify Your Success

(7)   Validate Your Success

(8)   Actualize Your Success

©Daymond E. Lavine

What makes this process even more interesting is the reason why the V-model is used. You see, there is symmetry in the model. There are eight steps you should complete for every goal of success you have. The four steps on the left side of the V-model are balanced by the four steps on the right of the V-model. Thus, "Conceive Your Success" is balanced with and is proven by "Actualize Your Success." "Plan Your Success" is balanced with and is proven by "Validate Your Success." "Define Your Success" is balanced with and is proven by "Verify Your Success." Finally, "Build Your Success" is balanced with and is proven by "Test Your Success." It will all become clear to you as you read through the pages of this book. I did my best to keep my language very clear so that you can adapt the principles herein to your life circumstances.

Section I

©Daymond E. Lavine

Success begins with each and every one of us identifying a problem or deficiency we have. We are often very clear about the problem we have, but we typically are not great at figuring out exactly what we must do to overcome it. I have a solution for this. You must firmly establish what the problem or deficiency is that you have. Then you must decompose your problem or deficiency into tasks you can perform to overcome it. So, what are those tasks you ask? Conceive, Plan, Define, and Build Your Success! In the Success System V^{SM}, I have identified this stage with the title "Problem Definition & Decomposition." It may sound complicated, but I will emphasize this stage may be as complicated as you choose to make it. Conceiving your notion of success, in most cases, is very easy; however, Planning and Defining Your Success can prove to be very daunting tasks. I have recommendations for your avoidance of spending too much time on these tasks. Once you get past these two tasks, Building Your Success becomes somewhat of an easy job. At that point, you will have a blueprint for success. Thus, it is extremely important that you Conceive, Plan, and Define Your Success with sound judgement, carefully thought-out rationale, a keen sense of self, and ideas about what you specifically and precisely desire in life. I will warn you in advance that if you are not seeking to become successful in accordance with your own grounded and centered desires, then you will still be successful using this model. However, you will not be *happy and successful*. Happiness comes with success when you achieve goals that uniquely resonate with who you are in the world—on your own terms. This is a key principle that I express and teach during my coaching and mentoring sessions with others.

Part 1
Conceive Your Success

©Daymond E. Lavine

When most people think of success, they visualize the things they want to have, the life they want to live, the places they think they should be or visit, the type of job they want, the kind of car they want to drive, and a host of other materialistic, superficial things. Don't get me wrong, it is wonderful to have goals and aspirations. Items that we acquire in life are often used to measure our success. However, when we look around and see all the things other people have and the lives they lead, many of us tend to think that when we acquire those same things or lead those same lifestyles, then we too will be successful. We get in this "keeping up with the Joneses" cycle that is not very good for us. However, the one good thing about this is that we are in the act of *Conceiving Our Success*. It is a very tricky act to engage in. One thing we always should keep in check during this act is making sure we are conceiving success that adequately aligns with our purpose and reason for being. I could write an entire new book all about finding your purpose in life, but I will keep my thoughts on this topic short and simple here. When you Conceive Your Success, make sure it is success based on your personal desires for happiness and joy. Evaluate your Concept of Success at its very inception to determine if you have desires based on what others have or want for you, or if you have desires spawned from personal goals you have for yourself . . . those that you know will make you happy. I cannot tell you exactly how you will figure this part out. However, I can tell you that you will *feel* exactly what is right for you. If you are a spiritual person like I am, then pray on it. Free yourself, free your mind, await the answer, and it will come.

Success Tip #1

Upon conception of your success, make sure it is not based on what others have or what others want for you; make sure it is based on your unique desires.

©Daymond E. Lavine

For the most part, Conceiving Your Success is an involuntary act. Our desires in life just emerge on whims. When they emerge, we should know how to adequately respond to them. My Success System V[SM] provides the solution.

Here are some examples of Conception of Success:

Example A

Anthony sits at his desk during his day job and wonders why he is so unhappy. A few days later, he realizes that he is unhappy because he is very limited in his capacity to do work to his fullest potential in his current work environment. He knows that he must change his work situation. He has a yearning to do what he thinks is his calling in life, and that is to function as a Chief Executive Officer of his own company, which provides software solutions for startup businesses. He is tired of being on software product development teams that lack fast-paced adaptation to cutting-edge trends in the marketplace. Anthony's idea of success is owning a software development company that allows him to generate a sustainable income while pushing himself to the limit. He wants to lead a company that stays on the cutting edge of software technology.

Example B

Stephanie is a manager at a retail department store. Her best friend Donna is a registered nurse. Donna recently traded her luxury sedan for a new luxury sports utility vehicle. Stephanie is happy for her friend, but she cannot help feeling as if she should be earning a greater income. She wants the same type of vehicle that Donna has, so Stephanie decides that she wants to be a registered nurse as well. She does not know the details of what it takes to become a registered nurse. In fact, the thought of being a registered nurse never entered her mind until the day Donna got her new luxury vehicle. Still, Stephanie knows that she can accomplish whatever she puts her mind to. She now wants to become a registered nurse, and nothing will stop her.

Example A and **Example B**, which describe the Conception of Success, are drastically different. One is spawned by a person's internal desire to achieve something better for himself. The other is based on a person's desire to achieve something very similar to what someone else has. We must always be careful to recognize when our desires for success come from our own personal goals to rise above issues and problems we face. This way, we may discern those moments when we conceive notions of success simply because someone else has something we do not have or is doing something that we want to do. Success comes in many shapes and forms. As human beings, we have the ability to create success for ourselves that may not be *the right success for us*. Before you embark on a path of using any of the information in this book to gain more success in your life, I suggest you base all your goals for success on personal assessments of your life and what you want out of it. Do not try to achieve something simply because you know someone else has done it. This road will surely lead you down a path to *unhappy* success.

Once you believe that you are ready to aptly accept your Concept of Success as something worthy of your pursuit, then write it down. If there are multiple Concepts of Success you have in mind, then write all of them down. It does not hurt to have multiple goals in life. Once you have them written down, then you can do the work to visualize them, prioritize them, and possibly de-conflict them, if necessary. There may be some Concepts of Success that simply cannot coexist in your life. Think about it: you cannot live in two dream locations at the same time, for instance. However, you can live in one part of the world during one part of the year, and another part of the world during another part of the year. You must be clear about something like this.

Your Concepts of Success may simply be statements at first; however, you should develop them very well before you start doing the rest of the tasks as dictated by the Success System V[SM]. Here is an example of a simple Concept of Success that many people have:

"I want to own my own business."

©Daymond E. Lavine

Lots of people have this Concept of Success in mind, but many of them do not know what it takes to own a business. For starters, take note of how ambiguous the statement is. This Concept of Success needs quite a bit more detail to guarantee some level of success for the conceiver. Here is an example of the same Concept of Success fleshed out to a level of detail that allows us to start using the Success System V[SM].

"I want to own an income-generating interior design business based in Dallas, Texas."

This Concept of Success presupposes the conceiver likes interior design. Perhaps he or she has always loved rearranging rooms or decorating homes for fun. Something inside of him or her forced him or her to consider interior design as a job. Thus, he or she identified the statement above as a Concept of Success and wrote it down. The next thing he or she should do is plan owning that interior design business. For simplicity, as you get through the remainder of this book, I will stick to this Concept of Success as the single thread that follows the entire Success System V[SM]. Put your seatbelt on! You will be going for a ride, but I promise to keep our journey as smooth as possible.

Part 2
Plan Your Success

©*Daymond E. Lavine*

I say this time and time again when I am working to coach or mentor people . . . you have got to have a plan! You cannot succeed without having a plan. If you do achieve what you consider to be some form of success without a plan, that means you have succeeded by chance or sheer luck. Success of that kind is messy. People who achieve things this way tend to stumble into success they cannot handle. That kind of success is often short-lived. Thus, I urge you to Plan Your Success. If you plan, then you will be ready for it when it comes. If you are not quite ready for what you receive from working your Success Plan, it's okay. By then, you will have the experience of Planning Your Success under your belt; so, you will subsequently have some level of knowledge to leverage for maintaining your success. That's a different topic for another day, perhaps another book.

Success Tip #2

Success may be based on whims, but it should always be planned!

©*Daymond E. Lavine*

Success Tip #3

Success planning is great! However, don't get caught in the minutiae of OVER PLANNING.

Planning is a very important step in the Success System V^{SM}. However, it can also be a pitfall. Some people immerse themselves in planning tasks they can never dig themselves out. They get caught up in the minutiae of over-detailing their plan. On the opposite end of the spectrum, some people do not know where to start with developing their plan for success. The outline provided below gives you a starting point for developing your plan for success. I also offer suggestions for how you should approach these sections of your Success Plan.

Success Plan Outline

(a)   Objectives

(b)   Resources

(c)   Schedule

(d)   Measurements

(e)   Expected Outcome(s)

Defining Your Success Plan Objectives . . .

In the *Objectives* section of your Success Plan, you will identify the primary statements that describe your Concept of Success. These statements will precisely state what it is exactly you hope to achieve as characteristics for your unique success. At this point in your planning process, you do not need to be extremely detailed with your Success Plan Objectives. They may be open-ended. Time frames do not even need to be established yet because you will define a timeline later. However, your Success Plan Objectives must contain characteristics of your unique success that may be objectively checked for attainment. Here is an example of a well-written Success Plan Objective.

©Daymond E. Lavine

Success Plan Objective 1

To own an interior design business that partners with residential and commercial companies in Dallas, Fort Worth, and the surrounding areas to provide contemporary interior design solutions

The Success Plan Objective above is very clear, and indeed, very achievable. You can see it is more precise in comparison to the original Concept of Success. The conceiver previously only desired to own a revenue-generating interior design business based in Dallas. Let's name the conceiver at this point. We will call her Janet. Janet has slept more since she originally spawned her Concept of Success. She has had some time to think more about several aspects of this Concept of Success. She thinks she can generate more income if she considers working the entire Dallas and Fort Worth regions. She also thinks she wants to focus on contemporary interior design, because she has been following interior design trends in the media. She knows that contemporary interior design is catching on with home owners and business owners like wildfire. She absolutely loves the trend and has even decorated her own home in accordance with eclectic contemporary trends. Janet receives compliments on her interior design skills all the time. So, she knows she would be great at it. Aside from this single objective above, Janet has thought of others she should capture as she moves forward with planning the establishment of her interior design business.

Success Plan Objective 2

To establish a website for my interior design business that effectively communicates a brand committed to contemporary interior design for both residential and commercial properties

Success Plan Objective 3

To establish an interior design portfolio that visually illustrates my interior design capabilities for both residential and commercial properties

Success Plan Objective 4

To have a brick-and-mortar interior design studio that features my contemporary design work

Again, the Success Plan Objectives above are achievable, and can be fully confirmed by observation. However, I want to also provide you some examples of objectives that are not so great. I want you to know the difference. See below.

Poor Objective Example 1

To own my own business that provides products that make people happy

Poor Objective Example 1 is bad because it is too vague. What kind of business does the person want to own? What products will the business provide? If you do not know what products you will provide, then how will you measure consumer happiness?

Poor Objective Example 2

To make more money so that I pay for things I want and need

Poor Objective Example 2 is bad because it is also very vague. How much more money does the person want to make? What are the things the person wants and needs? Knowing what you want and desire will define the amount of money you need to make to afford them. The person should define their wants and needs so he or she can define the monetary amounts needed to acquire them.

Poor Objective Example 3

To be a better person who shows people I am capable of doing great things

Poor Objective Example 3, once again, is vague. What will make the person better? What are the great things the person wants to do?

One of the many challenges we have as human beings, despite our capability of doing so much and accomplishing so many goals in life, is concisely defining our paths to success. This is one of our many human flaws, and I personally believe that it is a result of inaccurately knowing how much power we truly have. We must learn to be clear with ourselves, and we must learn to focus on our unique desires for success so that we may attain them. Then, we must document what we believe to be our means to our successful ends. There is no room for ambiguity when it comes to charting out our paths in life. As harsh as this may come across to you, I know that many of us are lazy when it comes to Defining Our Success. However, no one else will or can do this for us. It is completely in the hands of each individual—who is blessed with a sane mind and able body—to accurately define his or her successful path in life.

Success Tip #4

When pursuing your success, be clear and concise about your unique desires for success.

Identifying Your Success Plan Resources . . .

In the *Resources* section of your Success Plan, you will identify the resources needed for meeting the objectives you defined earlier. When a Concept of Success initially enters a person's mind, that person gets excited and invigorated, anxious to do something he or she has never done before. However, after a fleeting moment of ignorant bliss, which may last anywhere from five minutes to a few years, the person may then become upset, angry, frustrated, or sad because the thought of "resource needs" eventually enters his or her mind. Because this happens unexpectedly, it often comes as a bit of a shock. Typically, people are not ready for the reality check if they have not thought about it in the very beginning when a Concept of Success comes to mind. You may then hear that person say something like, "I thought I would be able to do that, but I gave up because I did not know it would be so difficult to do," or "Had I known it would take so much money to do that, then I would never have tried to do it to begin with." A lot of dreams get lost and a lot of goals go unaccomplished because many people fail to do the necessary investigation regarding the resources needed to make them successful. So, I am here to let you know you cannot bypass this very vital step in your success planning process.

Success Tip #5

Resources are needed for success, and you need to be aware of that and plan for it.

Janet has established four Success Plan Objectives to which she now needs to allocate resources. The following are some major resources she needs.

a. **A Business Infrastructure.** Janet will need to establish her business as a Limited Liability Corporation (LLC), Sole Proprietorship, Doing Business As (DBA) entity, or some other form of business in the State of Texas to ensure she provides goods and services in accordance with Texas State Law. Additionally, she should choose a business structure that suits her needs. For instance, DBAs are relatively easy to manage because they allow for you to conduct business using an established business name while you file business taxes using your already-established Social Security Number. However, all business liabilities fall directly on you. There are no laws that hold the owners of a DBA faultless should legal problems occur. However, LLCs function as a separate business entity with a newly given Employment Identification Number (EIN). Additionally, LLCs come with liability protection without all the formality and paperwork that S Corporations and C Corporations have. Depending on the business structure that Janet chooses, she may or may not need an EIN. After Janet adequately sets up her business to legitimately do business in the State of Texas and in accordance with the Internal Revenue Service (IRS), she will then be able to professionally seek out residential and commercial businesses for networking. However, there is something more she will need. Another resource she may want to consider is a preliminary database that she creates to store contact information for potential partnering residential and commercial businesses and their personnel.

b. **A Business Brand.** Janet will need to have a business brand established with an online presence; marketing materials such as business cards, letterhead and flyers; and other branded materials she believes will leave lasting impressions on her potential business associates. She will need to consider creating that potential partners database in tandem with establishing her business brand. Depending on her specific list of targeted businesses and business personnel, Janet will be able to create a brand that strategically attracts her potential partners as well as clientele.

c. **An Interior Design Portfolio.** Once Janet begins forming her relationships with potential partners and clients, they will ask to see her work. At that point, she will need to be able to show them some of the work she does. An online portfolio could meet this need. She will need to have plenty of great photos taken of her work. Thus, she will either hire a professional to do so, or acquire a camera that takes high-resolution photos.

d. **An Interior Design Studio.** A brick-and-mortar design studio is a very important resource for Janet to have, and may very well require a separate plan for establishment. However, for the sake of getting my point across in this book, Janet will need to at least consider the location for her facility, whether she will purchase or lease the space, and the costs she should consider for rent or securing a mortgage, insurance, and utilities.

e. **Startup Capital.** This is the most important resource of them all. Every entrepreneur needs money to start a business. There is no avoiding this. Every entrepreneur also has different amounts of startup capital at their disposal to utilize for getting their business off the ground. Janet will need to strongly consider the amount of money she has to get her interior design business up and running. Depending on her startup capital resources, she will launch her business either virtually or with a brick-and-mortar presence.

The Success Plan Resources Janet has just identified are only a start. As Janet delves more into her planning activities, she may identify other resources she needs to meet her Success Plan Objectives. Still, these identified objectives are great for getting her going so that she can subsequently begin scheduling the steps needed to eventually *actualize* her success.

When it comes to you figuring out what your exact resource needs are, always keep in mind they will fall into to one of three categories: (1) Time, (2) Money, and (3) Infrastructure, i.e. Tools and Spaces. Furthermore, more often than not, your resource needs will be multifaceted. Some of them will fall into more than one of these categories.

Success Tip #6

The resources you need for success will be categorized as time, money, and infrastructure, i.e. tools and spaces.

Defining Your Success Plan Schedule . . .

Today, many of us are extremely busy. We have a hard enough time just figuring out what we need to do from one day to the next. "What meetings do I have today, tomorrow, and the next day?" "What's for dinner?" "Where do my kids need to be for their after-school and weekend activities?" "When are my bills due?" "Did I walk the dog yet?" The list goes on and on, right? For this precise reason, you must begin writing things down. We have so many tools at our disposal to keep us on track that we can barely decide which one to use. Yet, most of us neglect to use any at all. We have calendars on our mobile phones. We have calendars on our computers too. We have apps for scheduling meetings, planning meetings, planning to plan meetings, and we can even conduct virtual meetings with an app these days. It is possible for us to link all of our electronic devices so that we may track and account for every single minute of our lives. So why is it so difficult for us to *schedule our success*? Is it because we do not have a crystal ball to see into the future and know what it takes to reach our goals? Perhaps we do not have the right education, or are we simply not as lucky as some other people around us? I prefer to think that each one of us is uniquely blessed. We all have qualities that provide us with advantages as well as disadvantages compared to the other people we observe in our daily lives. However, even though we do not have that crystal ball, I am offering you the opportunity to realize you have the next best thing. You have this book, a mind of your own, and the ability to chart out a set of strategic steps toward acquiring your desired success in life. You also have more control over your life than you may think. What you need to do is think of scheduling your success as a simple series of steps. Some of them will take short periods of time to complete. Others will take longer periods of time to complete. At any rate, once you develop a schedule for those steps, you will be able to constantly remind yourself where you are on your journey toward success. I will use Janet once again, for example.

Success Tip #7

Your path to success is a series of executable
steps that you are in control of executing.

One of the first resources we determined Janet needs for her success is an interior design business infrastructure. A business infrastructure is vital for success. Obviously, it should be one of the first things scheduled on her to-do list for success. Janet will need to figure out if she wants to set up an LLC or a DBA. She will need to do some research to figure out what business structure works best with her life plan. Let us say Janet will need **three weeks** to do all the extensive research and business networking necessary for her decision. Then, she will need **an additional week** to file all paperwork necessary to legally create her business presence. That information allows us to determine that upon completion of **Month 1**, Janet will have her business infrastructure set up within in the State of Texas and legally recognized by the IRS. Janet will also need a business bank account for her interior design business. This will only take her one day to complete. She will take care of this portion of setting up her business by making a trip to her bank of choice on a Saturday morning during **Month 1**.

The next thing Janet will need to think about is getting her business brand developed. Branding is very important for every business. However, one of the issues many entrepreneurs face is defining the correct scope of branding during the birthing stages of their businesses. I always mention to my branding clients that a business brand has many moving pieces. The brand encompasses the logo, the colors chosen to represent the business, the look and feel of the business website, the slogan and other descriptive statements chosen to market the business, the way the business owner represents his or her business, printed marketing material developed to represent and market the business, and so on.

Obviously, based on what I mentioned above about the many aspects of branding, defining a schedule for accomplishing branding tasks may prove to be overwhelming. Although it is relatively difficult, this is actually a defining moment. Every entrepreneur faces them, and every entrepreneur must learn to overcome defining moments to gain success. Defining moments are those when the entrepreneur should take in a deep breath, relax, and then state with utmost determination and belief, "Anything is possible!" Thus, Janet must identify the specific items for branding she would like to immediately acquire and implement. Those branding items include the following:

(a)   Acquiring a Logo

(b)   Defining a Tagline

(c)   Defining Mission, Purpose, Values, and Solutions Statements

(d)   Establishing a Website, with defined and edited content

(e)   Establishing Marketing Material, i.e. Virtual and Printed Flyers, Business Cards, Palm Cards, etc.

©Daymond E. Lavine

Success Tip #8

Tasks that seem overwhelming to us are defining moments we must overcome before we succeed.

Branding is an ongoing activity that occurs throughout the entire lifecycle of all businesses. Inherently, many people already know this. For that reason, they easily get wrapped into over-defining the branding for their businesses before actual business transactions occur. When you start a business, define the minimum, high-priority needs for your branding. This will greatly help you determine what the schedule will be for defining your fledgling business and brand.

Janet thinks it will take her approximately **6 months** to get her branding established for her interior design business. She already knows of a graphic design and branding company she might use to produce her business brand. Based on the initial meeting she had with the company, she and the company personnel estimated the entire process will take approximately **6 months**. Janet still has to fully flesh out a vision for her business. The branding company will translate her vision into the branded items mentioned previously. As those items are produced, Janet will have the opportunity to review in-work marketing collateral and provide feedback before any items are finalized.

Janet's interior design portfolio is another resource she will need for her interior design business. Janet has engaged in interior design projects in the past for friends and family. Unfortunately, she had never considered having an interior design portfolio until this moment. She now *needs* one. As she contemplates what she will do to get one in place, she gathers all the names and contact information of the friends and family she worked for. She decides she is going to offer them interior design services in return for acquiring fantastic, high-resolution photos that capture her design work at its best. These photos will serve as the basis for her interior design portfolio. After making all the necessary phone calls, Janet estimates it will take her approximately **6 months** to establish her interior design portfolio with the wonderful assistance of a photography student who has agreed to assist her. This 6-month period will overlap with the 6 months needed to establish her brand.

Other resources Janet needs to establish her interior design business are an interior design studio and startup capital. Janet has done some preliminary number crunching, and she

©Daymond E. Lavine

has already determined that owning a physical location for the business cannot happen during the first couple of years based on her current financial situation. Thus, she is foregoing looking for a physical location for her business until she can evaluate revenues generated from her new business. Afterwards, she will determine if a brick-and-mortar location is affordable.

As far as startup capital is concerned, Janet has funds on hand to get her business infrastructure implemented, her branding established, and her interior design portfolio in place on her new business website. When those items are established, her business will be fully operational!

Success Tip #9

A Simple Success Schedule and a Simple Success Budget allow you to measure your success as it relates to time and money!

©Daymond E. Lavine

Identifying Your Success Plan Measurements . . .

Measurement . . . that word seems so formal for simple planning, right? I think that is the reason why so many people fail to actually "measure" the steps they take along the way toward accomplishing their goals in life. The fact remains that we must always figure out a way to measure our success. This is the only way we will know if we are performing effective actions.

In the case of Janet's interior design business, there are two primary measurements to evaluate at this time. Those measurements are time and money as they relate to the resources she has previously identified. Thus, she has developed the following Simple Success Schedule and Simple Success Budget to help her meet her Success Plan Objectives.

	Jan	Feb	Mar	Apr	May	Jun	July	Aug	Sep	Oct	Nov	Dec
Establish Business Infrastructure			█									
Establish Business Brand				█	█	█	█	█	█			
Establish Marketing Database					█	█						
Establish Interior Design Portfolio					█	█	█	█	█	█		

Figure 2. Simple Success Schedule

Establish Business Infrastructure	$ 500
Establish Business Brand	$ 3000
Establish Interior Design Portfolio	$ 1000
TOTAL	$ 4500

Figure 3. Simple Success Budget

I must mention at this point that the above **Figure 2. Simple Success Schedule** and **Figure 3. Simple Success Budget** are only for Janet's initial launch of her interior design business. After the business launches, Janet will need to develop additional Success Plans for reaching her revenue-generation needs and obtaining her interior design studio. For now, remember that Janet has the following Success Objectives in her Success Plan:

Success Plan Objective 1

To own an interior design business that partners with residential and commercial companies in Dallas, Fort Worth, and the surrounding areas to provide contemporary interior design solutions

Success Plan Objective 2

To establish a website for my interior design business that effectively communicates a brand committed to contemporary interior design for both residential and commercial properties

Success Plan Objective 3

To establish an interior design portfolio that visually illustrates my interior design capabilities for both residential and commercial properties

Success Plan Objective 4

To have a brick-and-mortar interior design studio that features my contemporary design work

Success Tip #10

If you develop objectives for your success, then establishing expected outcomes for those objectives will keep you focused on exactly what you want.

©Daymond E. Lavine

Deriving Your Success Plan Expected Outcomes . . .

Expected outcomes are the results you are seeking to attain by meeting the Success Plan Objectives you established. They are typically statements that complement your Success Plan Objectives by providing the status that will fully substantiate you meeting those Success Plan Objectives.

Here are Janet's Expected Outcomes:

Expected Outcome Based on Meeting Success Plan Objective 1

My business formulation documents are adequately filed with the State of Texas and the IRS. A marketing database is in place for my partnership needs, and the database includes residential and commercial companies in Dallas, Fort Worth, and the surrounding areas. High-quality images of contemporary design work I have completed for both residential and commercial spaces are captured and saved for my marketing needs.

Expected Outcome Based on Meeting Success Plan Objective 2

I have an interior design business website that effectively captures the brand I want to promote for my interior design business. This website captures my mission, purpose, and solutions statements as well as my company logo and slogan.

Expected Outcome Based on Meeting Success Plan Objective 3

My interior design portfolio is housed on my brand-new interior design website. It captures my high-resolution images of completed contemporary interior design work. My online portfolio is available for use and future updating to feature any new interior design work I deem worthy of advertising on my website as time goes on.

Expected Outcome Based on Meeting Success Plan Objective 4

I have decided to forego looking for a physical location to house my interior design business at this time. I will spend the first three years evaluating my business's performance. If my income allows me to establish a brick-and-mortar location, then I will move forward with establishing my interior design studio.

©Daymond E. Lavine

Part 3
Define Your Success

What comes to your mind when you think of *Defining Your Success*? We have already covered conceiving success in **Section I Part 1**. We have even covered success planning in **Section I Part 2**. So, haven't we already covered defining success? No, we have not. Most people have very loose definitions of their personal success. Thus, it is no wonder why most people feel like success is a moving target. You cannot aptly attain what you want if do not fully comprehend *what it is* that you want first.

©Daymond E. Lavine

Success Tip #11

Success will not seem like a moving target to you if you fully define it for yourself first.

There is a technique that many companies use to "define" their success. Surprisingly, you can use it too! More often than not, you have not even bothered to conscientiously and consistently formalize attaining success for yourself. However, you have more than likely done it for a company or a business you have worked for. When I mention "Define Your Success" in this part of **Section I**, I precisely mean "Define Your **[Requirements for]** Success." Most importantly, when you "Define Your Success," do more—set yourself up to *demand your success*. Following the methodology established in my Success System V[SM] will allow you to do this.

If you grew up in a Christian home like I did, you had your very first encounter with demanding more of yourself with a set of rules found in the Holy Bible called the "Ten Commandments." In each one of those Ten Commandments, usage of the word "shall" is levied on humanity by the Creator, our God. You were taught to obey those Ten Commandments, or else, you would commit sin against Him. In essence, you learned early in life to demand more of yourself because God said to do so!

You see, we learned what it meant to be accountable for ourselves. Most people are fearful of committing sins. Thus, we try to abide by these Ten Commandments as closely as possible. Then, as we mature, we start to feel as though we are doing enough in life, simply because we are abiding by expectations set forth by these few Biblical demands of us.

When I embarked on my Systems Engineering career however, I learned that big business took a page out of religion's book . . . literally. As my career path unfolded in the engineering culture, my focus became well-established in the field of Requirements Definition. In every legally binding document that was negotiated between my company and internal customers or external customers, usage of the "shall-statement" was obligatory. In fact, I was trained to firmly identify every shall-statement as one that was legally binding. If that shall-statement was not adhered to by the party required to perform it, then there would be disappointing repercussions to suffer, ranging from loss of money or reputation to having to undergo trial proceedings for a breach of contract. Even when I left the engineering company and began working for a human resources technology firm, I noted that the contracts and proposals negotiated there had requirements statements within them as well. Usage of the word "shall" was not as rigorously enforced. However, at that time, I was already trained to mentally insert the word "shall" in statements I thought were important as negotiated terms of work for payment.

Success Tip #12

Create a contract with yourself to require yourself to do what must be done to be successful.

When you define your success, I recommend that you formalize your process just as big business does. I suggest you levy a contract with yourself, and *require* what must be done to reach the success you are seeking. I know this seems like a difficult task to perform, but once you get the hang of it, I promise you will be able to better define your Success Requirements as time goes on. Let us return to the case of Janet who desires to own an interior design business.

Janet has diligently finalized her Success Plan, but she is definitely not done. She now must focus on very specific requirements for several aspects pertaining to building her business. In actuality, Janet may spend days or weeks defining some vital requirements for her business. These requirements will need to be fulfilled to ensure the success of her building it. However, for the sake of simplicity, I will only focus on a few well-written requirements that complement all of her Success Plan Objectives. I will help you acquire a precise way of thinking so that you excel in your overall management of your personal success. We will treat the combination of Janet's Success Plan Objectives and the resources she needs to meet them as a Total Success System. Thus, we will define Success Requirements for Janet's *Total Success System*, otherwise known as her interior design business.

Let us first establish some high-level Success Requirements for the Total Success System. Again, Janet's interior design business is the Total Success System.

IDB1: My Interior Design Business shall have a legal infrastructure approved by the State of Texas and the IRS.

In the requirement above, "IDB1" is an acronym for "Interior Design Business (requirement) 1." For every requirement Janet defines, she will also establish a requirement identifier that represents the aspect of the Total Success System she is concerned with. For **IDB1**, why didn't I just write, "Janet's Interior Design Business shall have a legal infrastructure?" You see, in our everyday lives, we constantly operate in a state of ambiguity and chaos. Things are always changing around us. We have so many issues and occurrences to keep track of that we often contemplate ideas for overcoming problems, and then we quickly forget them. That is because we do not employ a formal system to utilize and implement them. When you begin to focus on exactly what you want out of life to be successful, you must not succumb to going with the flow of the vagueness around you. What does it mean to have "a legal infrastructure"? Well, in **IDB1**, Janet states precisely that. Janet's Interior Design Business must be legally filed with and approved by the State of Texas as well as recognized by the IRS. Every time Janet refers to this well-defined requirement, she will know that her business infrastructure must be formal, and this will not be forgotten.

IDB2: My Interior Design Business shall have a brand that includes a logo design incorporating the following items:

(a)   **The Business Name depicted in the design**

(b)   **The image of a House Window Pane depicted in the design**

(c)   **The color Royal Blue (#4169E1) or a hexadecimal variant of Royal Blue**

(d)   **Letter fonts that are derivatives of the Sans Serif category**

The requirement above presupposes that Janet spent some time thinking about a logo design she would like to see depicted throughout all her branding and marketing materials. The more objective and specific you are with your Success Requirements, the more successful you will be. Janet could have easily defined some subjectively specific requirements. For instance, Janet could have identified one of her logo design characteristics as being "a logo design that includes legible text for the business name." However, letter formats vary in legibility from one person to the next. What may be legible to her logo design producer may not be legible to her. Janet could have identified another subjective characteristic as such, "a logo design that is beautiful and feminine." Well, what does beautiful and feminine equate to visually? Again, what may be beautiful and feminine to her logo design producer may not be beautiful and feminine to her. If beauty and feminism are essential characteristics that must emanate from Janet's logo design, then she must identify objective elements that represent her ideas of beauty and feminism. For instance, "a logo design that incorporates butterfly graphics around the embedded window pane graphic." However, this design characteristic more than likely will not be established for Janet's logo design because her branding must communicate contemporary interior design. Branding for contemporary design is usually clean and crisp, with usage of minimalistic, geometric, and symmetric details. In general, less is more for contemporary design.

IDB3: My Interior Design Business shall have a virtual design portfolio embedded on the Interior Design Business Website.

The requirement above is simplistic. However, no more details are needed within the statement of this requirement. Either the virtual design portfolio is present on the interior design business website or not. But is anything missing? The answer is yes. More requirements will precipitate from this virtual design portfolio requirement. When Janet eventually focuses on the establishment of her virtual interior design portfolio, she will thus develop a subset of website portfolio specific requirements such as the following:

IDP1: My virtual Interior Design Portfolio shall include images taken by a professional photographer, which are saved with a high-resolution (hi-res) image quality of no less than 300 dpi.

IDP2: My virtual Interior Design Portfolio shall include hi-res images of design work with a contemporary theme for residential spaces.

IDP3: My virtual Interior Design Portfolio shall include hi-res images of design work with a contemporary theme for commercial spaces.

In the three requirements you have just read, the "IDP" acronym in the requirement identifiers represent "Interior Design Portfolio." Janet may develop more requirements for her virtual design portfolio depending on how detail-oriented she is regarding this aspect of her branding and online presence. However, she will want to consider defining only the requirements necessary for attaining the success she desires. After all, she will be personally accountable for ensuring all requirements she defines are met! Let's move on to Janet's interior design studio.

IDS1: My Interior Design Studio shall be located in the Dallas zip code 75207, the Dallas Design District.

IDS2: My Interior Design Studio shall have a lease rate of no more than $1.25 per square foot per month.

IDS3: My Interior Design Studio shall have a square footage of at least 3000 square feet.

In the three requirements above, the "IDS" acronym in the requirement identifiers represent "Interior Design Studio." Previously, we mentioned that Janet will forego securing a brick-and-mortar space for her interior design business. However, Janet loves to be well-prepared for her next moves in life. Thus, she defined these requirements. She already knows that she cannot afford space in the Dallas Design District for now. However, she aspires to afford a space that meets these requirements. I would like to mention here that although we may establish requirements for our success that we try our very best to meet, that does not mean we will not renegotiate our established Success Requirements in the future. You see, there may come a time when Janet is ready to establish a brick-and-mortar location, but her business revenues will

©Daymond E. Lavine

simply not allow her to set up shop in the Dallas Design District. Janet can then renegotiate her brick-and-mortar location requirement. She may eventually rewrite **IDS1** as follows:

IDS1: My Interior Design Studio shall be located in the Dallas zip code 75212, West Dallas along the Trinity River Corridor.

Success Tip #13

Just because you define your success at one point in time does not mean you cannot renegotiate it later due to circumstances.

Renegotiating your Success Requirements is a very important lesson to learn, and you must never forget it. Establishing requirements for your success is essential. Sticking to them is vital. However, getting paralyzed in a state of never being able to attain them is detrimental and destructive for your future! Do not kill your dreams by demanding more of yourself than you are able to give. Renegotiating your Success Requirements is okay!

Interfaces, The Forgotten Requirements

When we get to a point in life when we finally learn how to develop wonderful and well-defined requirements for our goals of success, we tend to focus our attention entirely on ourselves. For the most part, situations unfold just as we like, but there are still a few encounters that frustrate us. There are some hurdles we just cannot seem to overcome. We know the requirements are right. However, let me be clear here, we know the requirements are right for *our* specific needs. Yet, there are people, businesses, environments, and so forth for which we must also define requirements. We typically forget about these; and when we do, things do not go as expected. Allow me to explain this further.

Success Tip #14

To manage your own requirements for success,
you must define requirements for your interfaces
as well to accommodate your success.

Even though we have discussed a few requirements that Janet has defined for her interior design business, we have not discussed any requirements she has defined for vendors or service professionals she needs to engage. We know that Janet must establish a brand for her interior design business; and she cannot do this work alone. She needs to enlist the services of a brand development professional. Thus, she must also define requirements for this service provider based on her specific needs.

BDP1: The Brand Development Professional shall possess at least five years of professional experience developing brands for small businesses including the following services:

(a) **Logo Design**

(b) **Website Design**

(c) **Website Content Development**

(d) **Marketing Material Development, including Business Cards, Letterhead, and Flyers**

In the requirement above, "BDP" is an acronym for "Brand Development Professional." Let us consider the case that Janet did not think of this requirement before enlisting the services of a brand development professional. This would be an area of high risk and concern. Janet could find herself engaging the services of someone who lacks the experience she needs to capture the brand she has envisioned for herself. She could waste time and budget receiving unsatisfactory services and products. Then, she would need to move on from one service provider to the next aimlessly, because she would not have defined requirements to identify the right brand development professional for her.

The most important thing to remember about your interfaces is that they exist on multiple levels, and can be people, places, or things. At a higher level, in Janet's effort to establish her interior design business, she must work with a brand development professional. However, at a much lower level, Janet must get a website completed as part of her business branding efforts. Let's move on to Janet's new interior design business website.

> **IDW1: The Interior Design Website shall have a computer-monitor compatibility mode and a mobile-device (i.e. tablet, mobile phone) compatibility mode.**

> **IDW2: The Interior Design Website shall automatically switch to computer-monitor mode when accessed using a computer monitor.**

> **IDW3: The Interior Design Website shall automatically switch to mobile-device mode (i.e. tablet, mobile phone) when accessed using a mobile device.**

In the requirements above, the "IDW" acronym represents "Interior Design Website." Now that Janet has established these requirements for her website, she will be able to have effective communication with her brand development professional. Eventually, she will obtain a website that functions very well in the fast-paced world we currently live in. More frequently than ever before, websites are currently being accessed on mobile devices. Thus, Janet must have a website that meets **IDW1**, **IDW2**, and **IDW3**.

The interface requirements we covered above are just a few. Janet will have many more as she focuses on the people, places, and things she needs to be successful. You will too. Never forget that you must always manage your interfaces. The best way to do that is to define requirements for those interfaces first, and then make sure you enforce them!

©Daymond E. Lavine

Part 4
Build Your Success

So, we have gotten through what I consider the most crucial and difficult part of establishing success for nearly anything you would like to accomplish. We have covered the fact that your Concepts of Success and your Success Plan should be accompanied by Success Requirements for all important aspects of your desired success. When we define our Success Requirements in this manner, there is little chance of those Success Requirements being inconsistent with our Success Objectives. What we are doing, in essence, is staying firm in our commitment to success by formalizing our method for achieving it! With such a solid foundation established, Janet is now ready to *build* her success.

©Daymond E. Lavine

Success Tip #15

After you have defined your requirements for success, you have the "to-do" list in hand to create your success.

The requirements that Janet has established for her interior design business now serve as a to-do list. Every requirement that she has established must be fulfilled for her to be successful according to her desires. For instance, if Janet fulfills **IDB1** and **IDB2** in **Section I Part 3**, then she will have an interior design business that is legally established in the State of Texas and approved by the IRS. She will further have a logo design that meets the characteristics she desires. If Janet gets **IDP1**, **IDP2**, and **IDP3** fulfilled, then she will have an interior design portfolio that only includes 300 dpi, hi-res images for both residential and commercial spaces.

Establishing requirements for our success is how we demand our success! I have seen it time and time again whether I was working with my peers in Corporate America or with many of my own clients: failure to define requirements and defining poor requirements has the result of no success or success that is incongruent with our desires. If you have the right Success Requirements, then you will be well on the way to attaining the success you want. Simply tackle all the Success Requirements you have defined within the budget and time frame you have created for yourself. You will have created the blueprint for your success, so all you have to do thereafter is build it!

Keep in mind that even though you will have done your best to define requirements that you think you can achieve for all the high-priority aspects of the success you desire, there may be some requirements that prove to be unattainable. You only have a certain amount of time and budget. For these requirements, again, you must renegotiate them. You must consider the timelines you have identified in your plan and the budget you have on-hand, then renegotiate your requirements in a manner that meets your time and budget. Also consider revising your desires for your success if possible. If you can accept removing a requirement from your current scope of "things to do" while going down the path of requirements fulfillment, then, by all means, do that! Remember that Janet determined early in the employment of my Success System V^{SM} that she will not have enough budget to fulfill **IDS1**, **IDS2**, and **IDS3** for establishing a brick-and-mortar interior design studio. Thus, although she defined those requirements, she has removed them from her scope for now. Still, they have been archived for later reference.

Left Side V Wrap-Up!

Woo hoo! Now, we have traveled down the entire left side of the Success System V^{SM}. As a reminder, I am listing the following actions below that you should complete to solidify your path to success while engaging in the Problem Definition and Decomposition Stage of the Success System V^{SM}.

(1)   Conceive Your Success

(2)   Plan Your Success

(3)   Define Your Success

(4)   Build Your Success

At the very bottom of the Success System V^{SM}, between the success generation tasks of Building Your Success and Testing Your Success is the Solution Set Generation and Implementation stage. Just as the Success System V^{SM} illustrates, this stage of your success generation is very brief, but nonetheless, very important.

I think pictures tell a thousand words when they explain development processes. Thus, I have created **Figure 4** which follows.

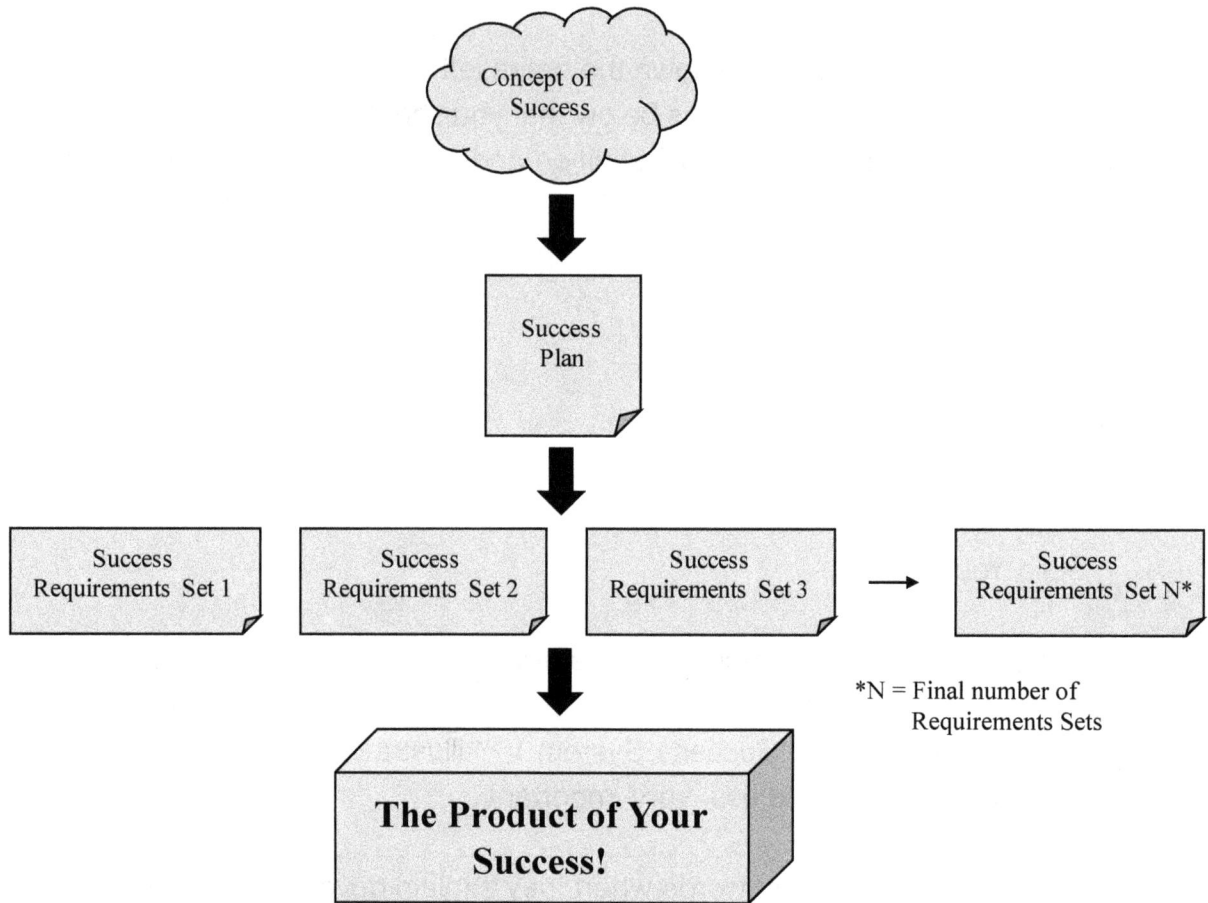

Figure 4. Success System V^{SM} Left-side Flow Down

©*Daymond E. Lavine*

Figure 4 depicts how the Concept of Success invokes a Success Plan. Once the Success Plan is developed, then Success Requirements are defined. After the Success Requirements are defined, then the Product of Success is built! All this activity occurs on the left side of the Success System V[SM]. Now that it is explained, the remainder of this book will focus on the right side of the Success System V[SM]. We will be moving on to **Section II**.

Section II

As many of us pursue our goals in life, we approach many decision points along the way, many of which we decide upon with very little direction or advice. We simply get to those decision points, and look at what others have done around us who have encountered similar situations. Perhaps we try to emulate other success stories we know about in news articles or have read about online. We move along in life, doing many things and making many mistakes along the way. However, wouldn't it be great to have a more definitive way to travel your path to success in life?

What my Success System V℠ does is allows you to treat your Success as a project you are in control of developing. The beauty of it is that it is a generic success model that allows you to completely set up the blueprint for your success, build it, and then evaluate it until it is actualized. In this portion of my book, I will now focus on moving up the path toward *Actualizing Your Success*. I will travel up the right side of the Success System V℠! Refer back to **Figure 1** if you need to be reminded of what that side of the Success System V℠ looks like.

Part 1
Test Your Success

After you have begun building your success, you will begin to have results from fulfilling your Success Requirements. This is wonderful news! You will be able to check your Success Requirements for attainment. Your goal is to make sure all the requirements you have identified for meeting the various aspects of your success are met. Thus, you will put them to the test!

Janet previously defined the following Success Requirement:

IDB1: My Interior Design Business shall have a legal infrastructure approved by the State of Texas and the Internal Revenue Service (IRS).

So, what is needed to make sure this Success Requirement is met? Janet has identified the following Success Verification Criteria (SVC) to ensure she meets this requirement:

IDB1 SVC: Acquisition of a stamped and approved Certificate of Formation for the Interior Design Business from the State of Texas. Acquisition of an official approval document from the IRS with the EIN for the interior design business.

Success Tip #16

Fulfilling your requirements for success allows
you to put your success to the test!

Janet has also defined the following Success Requirement regarding the brand development professional she needs to work with:

BDP1: The Brand Development Professional shall possess at least five years of professional experience developing brands for small businesses including the following services:

(a)   Logo Design

(b)   Website Design

(c)   Website Content Development

(d)   Marketing Material Development, including Business Cards, Letterhead, and Flyers

For this Success Requirement, Janet's Success Verification Criteria will be as follows:

BDP1 SVC: Acquisition of documentation and portfolio artifacts acquired online or directly from the brand development professional that captures the brand development professional's experience creating logos, websites, website content, and marketing materials. Acquisition of information acquired online or directly from the brand development professional that proves the brand development professional has been doing business for at least five years.

Do you see how it works? When you have solid requirements, you have a means to know exactly what you are looking for to constitute your success. If written well (i.e. objectively, clearly, and directly), you will be able to additionally capture objective artifacts and occurrences that ensure your Success Requirements have been met. In essence, you will be "testing" your requirements to objectively capture the information that supports their fulfillment.

©*Daymond E. Lavine*

Success Tip #17

When you establish solid requirements for your success, you create a means for knowing exactly what you desire regarding your success.

Part 2
Verify Your Success

©*Daymond E. Lavine*

In **Section II Part 1**, we discussed establishing the Success Verification Criteria that proves Success Requirements have be met. Thus, the act of you checking the Success Verification Criteria against all the Success Requirements you have created is *verifying your success*. At this point while invoking the Success System V^{SM}, your task becomes much easier, right? Yes it does! You will have already gone through your to-do list of Success Requirements, fulfilled them, and then captured your Success Verification Criteria artifacts for each of those requirements. Verifying your success is nothing more than checking your Success Requirements against the Success Verification Criteria artifacts you have saved and archived. You will be making sure that 100% of your requirements are fulfilled. Again, let me remind you that you may have defined some requirements early on that cannot be fulfilled due to time, budget, or sheer impossibility. If so, that means you have created *unrealistic requirements*. You must renegotiate them or get rid of them.

Success Tip #18

Verifying your success is nothing more than checking artifacts created while building your success against the criteria for meeting your requirements of success.

As far as verification of success is concerned, let us think of Janet once more, as an example.

> **IDB1: My Interior Design Business shall have a legal infrastructure approved by the State of Texas and the Internal Revenue Service (IRS).**

> **IDB1 SVC: Acquisition of a stamped and approved Certificate of Formation for the Interior Design Business from the State of Texas. Acquisition of an official approval document from the IRS with the EIN for the interior design business.**

For **IDB1** and **IDB1 SVC**, Janet will simply check to make sure she has received an approval document from the State of Texas and another from the IRS. She previously archived them. Thus, once she makes sure they are still in the archive folder she created in a secure location in cloud storage, she will denote that **IDB1** has been met.

> **BDP1: The Brand Development Professional shall possess at least five years of professional experience developing brands for small businesses including the following services:**

> **(a)   Logo Design**

> **(b)   Website Design**

> **(c)   Website Content Development**

> **(d)   Marketing Material Development, including Business Cards, Letterhead, and Flyers**

BDP1 SVC: Acquisition of documentation and portfolio artifacts acquired online or directly from the brand development professional that captures the brand development professional's experience creating logos, websites, website content, and marketing materials. Acquisition of information acquired online or directly from the brand development professional that proves the brand development professional has been doing business for at least five years.

For **BDP1** and **BDP1 SVC**, Janet will simple check to make sure that she has reviewed her brand development professional's credentials and examples of his past work. Once she does this, she will mark **BDP1** as "fulfilled" and move on. This will be Janet's standard process for making sure all her Success Requirements have been met.

In your case, due to schedule pressure, you may relax your capture and archiving of Success Verification Criteria artifacts. Janet may have the same problem. For example, once she checks the credentials of her brand development professional, she may simply mark **BDP1** as "fulfilled." This is fine. However, I do encourage you to archive whatever artifacts you think will cause you grief if you cannot easily put your hands on them any time in the future. For instance, Janet should definitely archive those approval documents from the State of Texas and the IRS. They prove her interior design business is legally formed. You see, she is not the only interested party concerned with those documents. The IRS, bank institutions, business loan agents, and other business professionals may request copies of them on later dates depending on other goals she has for her business.

©Daymond E. Lavine

Part 3
Validate Your Success

Now we are on a roll! We have gotten through the idea that 100% of your Success Requirements must been met. Thus, we will need to take one more step to *validate your success*. Success verification and validation may be confusing. You may be tempted use the term *verification* and *validation* interchangeably; however, you **should not**. Let me explain. The process of verifying your Success Requirements involve you checking all your Success Verification Criteria artifacts to ensure they fulfill your Success Requirements. However, the process of validating your success is ensuring all aspects of your success as dictated by your Success Plan Objectives are satisfactory to you. This a huge difference!

We have previously covered the fact that Janet developed a set of requirements for her interior design business. The reason why she developed those requirements is that she wanted an interior design business in accordance with her Success Plan. So, once Janet has ensured all her Success Requirements have been met for her Total Success System (i.e. her interior design business), she will additionally make sure her interior design business is satisfactory to her, at all levels. Let us use Janet's interior design portfolio for instance. When she finishes fulfilling the Success Requirements for it, she will then treat her interior design portfolio as a Success System within her Total Success System. (Remember that Janet's interior design business, as a whole, is considered the Total Success System; but it has multiple aspects within it with their own associated Success Requirements.) Janet can visit her interior design portfolio online for herself and perform several actions and clicks to make sure nothing unsatisfactory occurs. When she does this, she may encounter some functionality issues. If so, this means a Success Requirement was missing when her brand development professional built her virtual interior design portfolio. Janet will need to contact her brand development professional and discuss the possibility of fixing the issue. This is an issue that could impact her schedule and budget. It also serves as an example of the consequences that could arise when we forget to define and implement Success Requirements.

©*Daymond E. Lavine*

Success Tip #19

When you establish solid requirements for your success, you create a means for knowing exactly what you desire regarding your success.

Part 4
Actualize Your Success

Our hopes and dreams can become reality! *Actualizing Your Success* is nothing more than bringing the concepts of your success to life. The term speaks for itself. Thankfully, we thoroughly covered the Concept of Success in **Section I Part 1**. We determined that it should be based on personal desires near and dear to our hearts opposed to success based on what others have or want for us. When you establish a basis for your unique desires for success, and follow the Success System V^{SM}, then you actualize success that brings you happiness and joy!

Make sure you utilize all we have covered so far. Be certain to refer to Janet's actions as examples for manifesting your own Total Success System. Sure, Janet wanted to specifically establish an interior design business, but you may want to do or build something else that you consider to be successful. Perhaps you want to learn a new skill. Well, go and plan it out; then, define the Success Requirements needed for you to gain that knowledge. Perhaps you want to build a business of your own that you consider to be successful, just as Janet did. Again, plan that. Then, go and define the Success Requirements for that as well.

When most people think of success, they think of getting rich or famous. However, I want to make sure you realize that success happens on many levels. Success is scalable, and it totally depends on the individual who conceives it. This is my moment to stress to you this personal pledge you should make for yourself. **Do not let anyone or anything detour you from attaining your unique success.** Another way I can say this is, do not let anyone or anything detour you from attaining **your best life**.

Success Tip #20

Conceiving your success begins a path that must be fully traveled before actualizing your success.

©*Daymond E. Lavine*

Right Side Wrap-Up!

Woo hoo! Now, we have traveled up the entire right side of the Success System V[SM]. As a reminder, I am listing the following actions below that you should complete to solidify your path to success while engaging in the Problem Assessment and Minimization stage of the Success System V[SM].

(1)   Test Your Success

(2)   Verify Your Success

(3)   Validate Your Success

(4)   Actualize Your Success

When you utilize the Success System V[SM] as defined and explained in the context of this book, you will be successful. You will be successful because you will have a well-planned, well-executed blueprint and model for success! That is what the Success System V[SM] is all about. It organizes all the chaos we have going on in our minds regarding our dreams of success into tactical actions we can implement and substantiate!

Success Tip #21

The Success System V^{SM} allows you to attain success by arming you with a well-planned and well-executed blueprint and model for success!

©Daymond E. Lavine

Conclusion

I have often been viewed as the problem-solver in nearly every business arrangement I have encountered. Consistently throughout my entire Systems Engineering career, I have overcome some very tough problems. I have done so in a very well-organized and executable manner. Because I branded myself as a great problem-solver, even among my entrepreneurial business clients, I have become appreciated as a doer and service provider who helps people get things done and meet their goals. However, it does not stop there. The very same techniques I identified in the pages of this book using the Success System V^{SM} are the very same ones I employ to create my own success. I know these methods work. Sure, reaching our goals of success will always depend on some trial and error. However, efficiency is very important as well. My Success System V^{SM} creates an efficient way to reach personal goals for success!

If this book has found its way into your hands or in front of your eyes, I believe it is meant to be there. Nothing happens by chance in this life. You can choose to believe it or not; however, I know this in my heart and in my mind. My spiritual beliefs will not allow me to think otherwise. For this reason, I urge you do something very important: do some introspection before you embark on following the Success System V^{SM} for yourself. You see, I say this because of my own experience with it.

What is acquired success that does not truly make you happy? I went down this path. I pursued goals of success for the sake of making money or attaining materialistic things. Then, when I attained that success, I became somewhat unhappy. This was unsettling for me because I then became a slave to the very same success I had created for myself. There was a lesson for me in that experience. I had to learn that all success is not "happy" and "joyful" success. Thus, I urge you to make sure you do not embark on your journey toward success in a manner that does not truly resonate with you. Make sure that you become successful while remaining who you were born to be.

Nothing happens by chance; I must say this again. With the creative mind I have, I know I was indeed meant to become a successful Systems Engineer. However, I also know that I was born to bring this book to life. I exist to inspire, and I have managed to allow you and so many of my peers understand exactly how my mind works. Now you know how I get through my tough problems in life and attain my personal success. Please study the pages of this book diligently and with conviction. Oh, and be careful what you wish for after reading it; because if you follow my Success System V^{SM}, you just might get it!

XOXO DAYMOND

TECHNICALLY INCLINED. CREATIVELY DRIVEN.
"I Exist to Inspire"

#deliberatesuccess #wired2inspire

Visit me online at DaymondCo.com

©Daymond E. Lavine

NOTES:

NOTES:

A SIMPLE
Systems Engineering
Guide for Success

D. & CO.

DAYMOND E. LAVINE

This page is intentionally left blank.

www.ingramcontent.com/pod-product-compliance
Lightning Source LLC
Chambersburg PA
CBHW050639150426
42813CB00054B/1113